Getting Ready for
High-Stakes
Assessments

INCLUDES:
- Standards Practice
- Beginning-, Middle-, and End-of-Year Benchmark Tests with Performance Tasks
- Year-End Performance Assessment Task

ISBN 978-0-544-60193-2

2 3 4 5 6 7 8 9 10   1689   23 22 21 20 19 18 17 16 15

4500530633   A B C D E F G

Contents

High-Stakes Assessment Item Formats

The various high-stakes assessments contain item types beyond the traditional multiple-choice format, which allows for a more robust assessment of children's understanding of concepts.

Most high-stakes assessments will be administered via computers; and *Getting Ready for High-Stakes Assessments* presents items in formats similar to what you will see on the tests. The following information is provided to help familiarize you with these different types of items. Each item type is identified on pages (vii–viii). The examples will introduce you to the item types.

The following explanations are provided to guide you in answering the questions. These pages (v–vi) describe the most common item types. You may find other types on some tests.

Example 1 Choose numbers less than a given number.

More Than One Correct Choice

This type of item looks like a traditional multiple-choice item, but it asks you to choose all of something. When the item asks you to find all, look for more than one correct choice. Carefully look at each choice and mark it if it is a correct answer.

Example 2 Choose tens and ones to describe a number.

Choose From a List

Sometimes when you take a test on a computer, you will have to select a word, number, or symbol from a drop-down list. The *Getting Ready for High-Stakes Assessments* tests show a list and ask you to choose the correct answer. Make your choice by circling the correct answer. There will only be one choice that is correct.

Example 3 Sort numbers into groups for greater than or less than a given number.

Sorting

You may be asked to sort something into categories. These items will present numbers, words, or equations on rectangular "tiles." The directions will ask you to write each of the items in the box that tells about it.

Sometimes you may write the same number or word in more than one box. For example, if you need to sort quadrilaterals by category, a square could be in a box labeled rectangle and another box labeled rhombus.

Example 4 Order numbers from least to greatest.

Use Given Numbers in the Answer

You may also see numbers and symbols on tiles when you are asked to write an equation or answer a question using only numbers. You should use the given numbers to write the answer to the problem. Sometimes there will be extra numbers. You may also need to use each number more than once.

Example 5 Match related facts.

Matching

Some items will ask you to match equivalent values or other related items. The directions will specify what you should match. There will be dots to guide you in drawing lines. The matching may be between columns or rows.

Item Types:

Example 1

More Than One Correct Choice

Fill in the bubble next to all the correct answers.

Choose all the numbers less than 25.

○ 32

○ 24

○ 52

○ 17

○ 61

Example 2

Choose From a List

Circle the words.

What is another way to write 24?

2 | tens
ones

4 | tens
ones

Example 3

Sorting

Copy the numbers in the right box.

Write each number in the box that tells about it.

| 33 | 46 | 72 | 97 |

Less than 50	Greater than 50

Example 4

Use the Numbers

Write the numbers.

Write the numbers in order from least to greatest.

| 18 | 12 | 21 | 8 |

_____ _____ _____ _____

Example 5

Matching

Draw lines to match.

Match the related facts.

3 + 2 = 5 •   • 9 − 6 = 3

8 − 2 = 6 •   • 2 + 3 = 5

3 + 6 = 9 •   • 2 + 7 = 9

9 − 7 = 2 •   • 8 − 6 = 2

1. Jerry picked 16 red apples. He picked 27 green apples. How many apples did Jerry pick?

Label the bar model. Write a number sentence with a ▓ for the missing number. Solve.

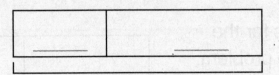

_____

_____   _____ apples

---

2. George collected 36 oak leaves. He collected 41 maples leaves. Show how you can break apart the addends to show how many leaves George collected.

36 ⟶   _____ + _____

+ 41 ⟶   _____ + _____

     _____ + _____ = _____

---

3. Ben has 37 pennies. Rachel has 22 pennies. How many pennies do they have?

_____ pennies

GO ON ▶

4. Kent's book has 64 pages. Kent reads 27 pages in the morning. He reads 18 more pages that night. How many pages does Kent need to read to finish the book?

Complete the bar models for the steps you do to solve the problem.

_____ more pages

5. A boat can hold 45 people. 27 more people can fit on the boat. How many people are on the boat now?

Write a number sentence for the problem. Use a ■ for the missing number. Then solve.

_____

There are _____ people on the boat.

6. John has 16 rocks. He gives 7 rocks to his cousin. How many rocks does John have now? Write a number sentence for the problem. Use ■ for the missing number. Then solve.

_____

_____ rocks

I. Fill in the bubble next to all the doubles facts you could use to find the sum of $4 + 5$.

○ $2 + 2$

○ $3 + 3$

○ $4 + 4$

○ $5 + 5$

2. Which number sentence has the same difference as $15 - 8 = \blacksquare$?

○ $10 - 1 = \blacksquare$

○ $10 - 2 = \blacksquare$

○ $10 - 3 = \blacksquare$

○ $10 - 4 = \blacksquare$

3. There are 8 large plates and 7 small plates on a table. How many plates are on the table? Write the number sentence. Show how you can make a ten to find the sum. Write the sum.

$$8 + 7 = \underline{\qquad}$$

$$10 + \underline{\qquad} = \underline{\qquad}$$

GO ON

Name _____

4. Ling sees 13 birds in a tree
and 5 birds on the ground.
How many more birds does Ling
see in the tree than on the ground?
Draw a picture and write a number
sentence to solve.

_____

_____ more birds

5. There are 9 oranges in a bag. Mr. Johnson
puts 4 more oranges in the bag. Complete
the addition sentence to find how many
oranges are in the bag now.

_____ + _____ = _____

_____ oranges

6. Use the numbers on the tiles to write the differences.
Then write the next fact in the pattern.

| 6 | 7 | 8 | 9 |

11 − 3 = _____   10 − 3 = _____

11 − 4 = _____   11 − 3 = _____

11 − 5 = _____   12 − 3 = _____

_____   _____

(STOP)

Name _____

Practice Test

2.OA.C.3
Work with equal groups of objects to gain foundations for multiplication.

1. Choose the ten frame that shows an even number.

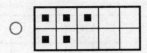

2. Write an even number between 3 and 12.
   Draw a picture and then write a sentence
   to explain why it is an even number.

_____

_____

3. Mary has an even number of shells and an odd
   number of rocks. Choose all the groups of shells
   and rocks that could belong to Mary.

   ○ 4 shells and 8 rocks   ○ 9 shells and 6 rocks

   ○ 8 shells and 3 rocks   ○ 4 shells and 7 rocks

GO ON ➡

Name _____

4. Choose the ten frame that shows an
odd number.

○    ○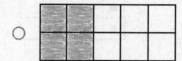

○    ○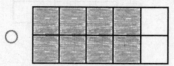

5. Write an even number between 7 and 16.
Draw a picture and then write a sentence
to explain why it is an odd number.

_____

_____

6. Ted has an even number of yellow markers and an
odd number of green markers. Choose all the
groups of markers that could belong to Ted.

○ 8 yellow markers and 3 green markers

○ 3 yellow markers and 6 green markers

○ 4 yellow markers and 2 green markers

○ 6 yellow markers and 7 green markers

Practice Test

2.OA.C.4
Work with equal groups of objects to gain foundations for multiplication.

1. Find the number of shapes in each row.

3 rows of _____

Complete the addition sentence to find the total.

_____ + _____ + _____ = _____

---

2. Max and his sister each have 4 playing cards.
Draw a picture to show the groups of cards.

How many playing cards do they have?

_____ playing cards

---

3. Find the number of shapes in each row.

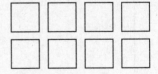

2 rows of _____

Complete the addition sentence to find the total.

_____ + _____ = _____

GO ON ➡

Name _____

4. Write how many in each row and in each column.

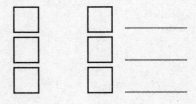

_____

_____

_____

_____   _____

---

5. Fill in the bubble next to all of the equations that match the array above.

○ $2 + 2 = 4$

○ $3 + 3 = 6$

○ $2 + 2 + 2 = 6$

○ $3 + 3 + 3 = 9$

---

6. Tanya and 2 friends put rocks on the table. Each person put 2 rocks on the table. Draw a picture to show the groups of rocks.

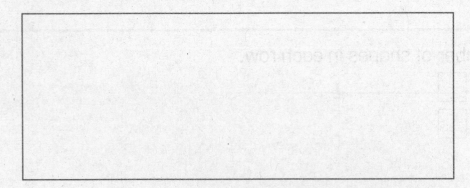

How many rocks did they put on the table?

_____ rocks

1. Sonya has 140 beads. How many bags of 10 beads does she need so that she will have 200 beads in all?

   _____ bags of beads

2. Choose all the numbers that have the digit 8 in the tens place.

   ○ 148

   ○ 387

   ○ 836

   ○ 881

3. Draw 139 using hundred boxes, ten sticks, and circles. Write the number name. Then write the number in expanded form.

Name _____

4. Robin has 180 stickers. How many pages of
10 stickers does she need so that she will have
200 stickers in all?

_____ pages of stickers

---

5. Choose all the numbers that have the digit 2
in the tens place.

○ 721

○ 142

○ 425

○ 239

---

6. Count the hundreds, tens, and ones.
Write the total.

☐ ☐ ☐ oooo
            oooo

_____   _____   _____   Total _____

Hundreds   Tens   Ones

1.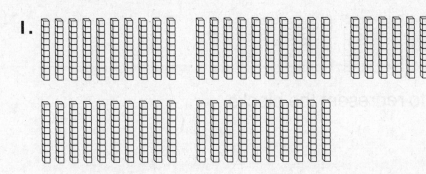

Choose all the ways to represent the blocks.

○ 50 hundreds

○ 50 tens

○ 5 hundreds

○ 5 tens

2. Pencils are sold in boxes of 10 pencils.
Mr. Lee needs 100 pencils. He has 40 pencils.
How many boxes of 10 pencils should he buy?

_____ boxes of 10 pencils

Draw a picture to explain your answer.

Name _____

3.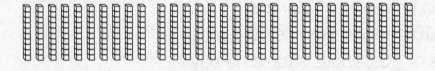

Choose all the ways to represent the blocks.

○ 3 hundreds

○ 30 ones

○ 30 hundreds

○ 30 tens

---

4. Write how many tens. Write how many hundreds.
Write the number.

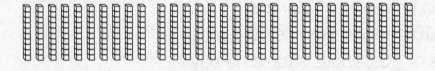

_____ tens

_____ hundreds

---

5. Write how many tens. Write how many hundreds.
Write the number.

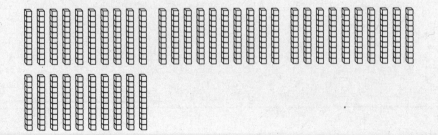

_____ tens

_____ hundreds

_____

Name ____________________

Practice Test
2.NBT.A.2
Understand place value.

1. Jen starts at 280 and counts by tens.
What are the next 6 numbers Jen will say?

280, ______, ______, ______, ______, ______, ______

---

2. Skip count by 5s.

35 ______ ______ ______ ______ ______ 65

---

3. Write the numbers that complete each counting pattern.

25, ______, 35, 40, 45

10, 20, 30, ______, 50

50, 60, 70, 80, ______

70, 75, ______, 85, 90

GO ON

Name _____

4. Jeff starts at 190 and counts by tens. What are the next 6 numbers Jeff will say?

190, _____, _____, _____, _____, _____, _____

5. Choose the ways that show counting by tens.

○ 550 560 570 580 590 600 610 620 630

○ 200 300 400 500 600 700 800 900 1,000

○ 650 651 652 653 654 655 656 657 658

○ 210 220 230 240 250 260 270 280 290

○ 170 180 190 200 210 220 230 240 250

6. Choose the counting pattern that shows counting by fives.

○ 76, 77, 78, 79, 80

○ 20, 30, 40, 50, 60

○ 70, 75, 80, 85, 90

○ 40, 42, 44, 46, 48

1. What is the value of the digit 1 in the
   number 41?

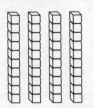

   _____

2. Draw a picture to show the number 52.

   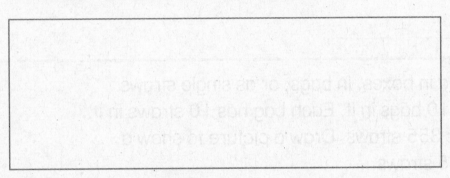

   Describe the number 52 in two ways.

   [2 5] tens   [2 5] ones

   _____ + _____

3. Dan has a favorite number.
   His number has the digit 8 in
   the ones place and the digit 5
   in the tens place. What is
   Dan's favorite number?

   _____

   GO ON ➡

Name _____

4. It's 154 days until Jeff's birthday. Write the number of days in words.

_____

Show the number in two other ways.

Hundreds	Tens	Ones

_____ + _____ + _____

5. Straws are sold in boxes, in bags, or as single straws. Each box has 10 bags in it. Each bag has 10 straws in it. Mr. Tan needs 355 straws. Draw a picture to show a way to buy 355 straws.

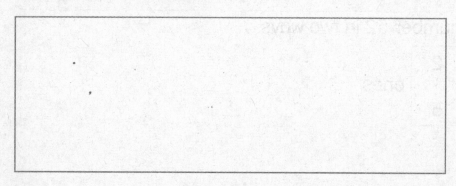

How many boxes, bags, and single straws did you show?

_____

_____

6. Terry has 164 marbles.

Write the number in words.

_____

1. Fill in the bubble next to all the comparisons that are true.

○ 343 < 328

○ 705 > 699

○ 691 > 706

○ 115 < 120

---

2. Jill and Ed collect postcards. Jill has 124 postcards. Ed has 131 postcards. Who has more postcards? _____

Jill gets 10 more postcards. Ed gets 5 more postcards. Who has more postcards now? _____

Draw quick pictures to show how many postcards Jill and Ed have now.

Jill's postcards	Ed's postcards

---

3. Write a symbol from a tile to compare numbers.

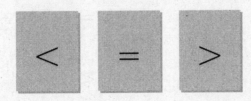

143 ◯ 143

GO ON ➡

Name _____

4. Write the symbols that make each comparison true. Choose >, <, or =.

787 _____ 769

405 _____ 399

396 _____ 402

128 _____ 131

5. Dan and Hannah collect toy cars. Dan has 132 cars. Hannah has 138 cars. Who has more cars? _____

Dan gets 10 more cars. Hannah gets 3 more cars. Who has more cars now? _____

Draw quick pictures to show how many cars Dan and Hannah have now.

Dan's Cars	Hannah's Cars

6. Match the pairs of numbers to =, <, or >.

183 ( ) 138   . =

182 ( ) 208   . <

947 ( ) 947   . >

1. Without finding the sums, choose all the pairs of addends that have sums greater than 100.

○ 77 + 12

○ 49 + 72

○ 16 + 95

○ 47 + 46

Explain how you decided which pairs have a sum greater than 100.

_____

_____

2. Which pair of numbers do you need to regroup to subtract?

○ 55 − 24

○ 39 − 18

○ 30 − 19

○ 73 − 63

3. Use the number line. Count up to find the difference.

53 − 46 = _____

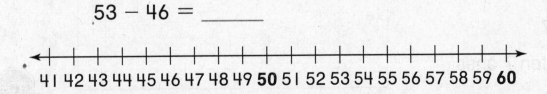

41 42 43 44 45 46 47 48 49 **50** 51 52 53 54 55 56 57 58 59 **60**

GO ON →

4. What is 46 − 28? Use the numbers on the tiles to rewrite the subtraction problem. Then find the difference.

| 18 | 28 | 46 | 74 |

```
  [   ]
−
  [   ]
─────
  [   ]
```

Regroup if you need to. Write the difference.

5.

Tens	Ones
[ ]	[ ]
4	6
− 1	9

6.

Tens	Ones
[ ]	[ ]
7	5
− 2	5

---

7. Subtract 18 from 35. Draw to show the regrouping. Fill in the bubbles next to all the ways to write the difference.

○ 7 tens 1 one

○ 17

○ 1 ten 7 ones

○ 71

Practice Test

2.NBT.B.6
Use place value understanding and
properties of operations to add and subtract.

1. There are 18 girls in the chorus. There are 9 boys.
   How many children are there in the chorus?

   Draw quick pictures to solve. Write the sum.

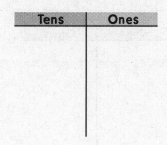

Tens	Ones

_____ children

Did you regroup to find the answer? Explain.

_____

_____

2. On four days of this week, Madelyn practiced her violin
   43 minutes, 32 minutes, 16 minutes, and 25 minutes.

   Write a number sentence to show the number of
   minutes Madelyn practiced her violin.

   _____

3. Olivia sees 16 crabs at the beach. Then she
   sees 7 more. Choose all the ways you can
   use to find how many crabs Olivia sees.

   ○ 16 − 7

   ○ 16 + 7

   ○ 16 + 4 + 3

GO ON ➡

Name _____

4. Lauren sees 14 birds. Her friend sees 7 birds.
How many birds do Lauren and her friend see?
Draw quick pictures to solve. Write the sum.

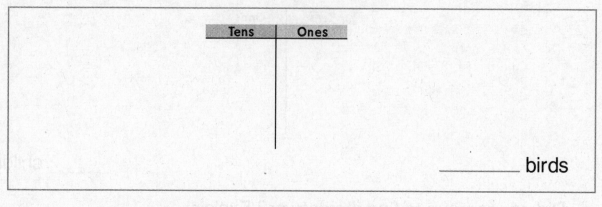

Tens	Ones

_____ birds

Did you regroup to find the answer? Explain.

_____

_____

5. Leah put 21 white marbles, 31 black marbles, and
7 blue marbles in a bag. Then her sister added
19 yellow marbles.

Write a number sentence to show the number
of marbles in the bag.

_____

6. Leslie finds 24 paper clips in her desk. She finds
8 more paper clips in her pencil box. Choose all
the ways you can use to find how many paper
clips Leslie has in all.

○ 24 + 8

○ 24 − 8

○ 24 + 6 + 2

Name _____

Practice Test

2.NBT.B.7
*Use place value understanding and
properties of operations to add and subtract.*

1. At the store, there are 863 rocks and
shells. There are 121 rocks. How many
shells are there? Circle the number that
makes the sentence true.

There are | 642 / 732 / 742 | shells.

2. A bird watcher counted 163 white birds and
185 black birds. How many birds did she count?

$$163 \longrightarrow 100 + 60 + 3$$
$$+ 185 \longrightarrow + 100 + 80 + 5$$

Select one number from each column to
solve the problem.

Hundreds	Tens	Ones
○ 2	○ 4	○ 6
○ 3	○ 5	○ 7
○ 4	○ 6	○ 8

GO ON

3. A farmer has 305 sheep. She moves 263 sheep into a field.

Choose all the sentences that describe how to find how many sheep are left.

○ Regroup I hundred as I0 tens.

○ Regroup I ten as I0 ones.

○ Subtract 6 tens from I0 tens.

---

4. Use the numbers on the tiles to solve the problem.

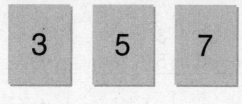

Describe how you solved the problem.

_____

_____

_____

_____

STOP

Practice Test
2.NBT.B.8
Use place value understanding and properties of operations to add and subtract.

1. A store has 263 board games. It has 100 fewer puzzles than board games. The store has 10 more action figures than puzzles. Write the number of each.

_____   _____   _____
   board games   puzzles   action figures

2. Write the next number in each counting pattern.

162, 262, 362, 462, _____

347, 357, 367, 377, _____

609, 619, 629, 639, _____

3. Use the clues to answer the question.

• Shawn counts 213 cars.
• Maria counts 100 fewer cars than Shawn.
• Jayden counts 10 more cars than Maria.

How many cars does Jayden count?   _____ cars

4. Sanjo has 348 marbles. Harry has 100 fewer marbles than Sanjo. Ari has 10 more marbles than Harry. Write the number of marbles each child has.

_____   _____   _____
 Sanjo   Ari   Harry

5. Rick has 10 more crayons than Lori. Lori has 136 crayons. Tom has 10 fewer crayons than Rick. How many crayons does each child have?

Rick: _____ crayons

Tom: _____ crayons

Lori: _____ crayons

6. Rico has 235 stickers. Gabby has 100 more stickers than Rico. Thomas has 10 fewer stickers than Gabby. Write the number of stickers each child has.

_____   _____   _____
 Rico   Gabby   Thomas

STOP

Practice Test
2.NBT.B.9
Use place value understanding and
properties of operations to add and subtract.

1. Draw a quick picture to solve. Write the difference.

Tens	Ones
8	I
− 5	6

Tens	Ones

Explain what you did to find the difference.

_____

_____

_____

2. Diana has 196 marbles. She gives away 42 of her
marbles to her sister. What step should Diana
follow first to find out how many marbles she has now?

○ Ungroup 6 tens as 5 tens 10 ones.

○ Subtract 2 ones from 9 tens.

○ Subtract 2 ones from 6 ones.

○ Subtract 4 ones from 9 tens.

GO ON

Name _____

3. Subtract 46 from 73. Explain all the steps you use.

$$\begin{array}{r} 73 \\ -46 \\ \hline \end{array}$$

4. Draw a quick picture to solve. Write the difference.

Tens	Ones
□	□
6	2
− 2	5

Tens	Ones

Explain what you did to find the difference.

_____

_____

_____

STOP

1. Owen wants to measure the length of a chalkboard.

Circle the best choice of tool.

tiles	inch ruler	yardstick

Explain your choice of tool.

_____

_____

2. Zach uses tiles to measure a straw. Each tile is 1 inch long. Zach says the straw is 4 inches long. Is he correct? Explain.

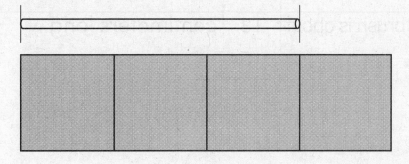

_____

_____

3. Josh wants to measures the distance around
a soccer ball.

Circle the best choice of tool.

inch ruler   yardstick   measuring tape

Explain your choice of tool.

_____

_____

4. Measure the length of the paintbrush to the
nearest centimeter. Circle the number in the
box that makes the sentence true.

The paintbrush is about
| 12 |
| 13 |
| 15 |
centimeters long.

1. Write the word that makes the sentence true.

centimeters	meters

A pencil is 16 _____ long.

A swimming pool is 50 _____ long.

A sidewalk is 2 _____ wide.

A computer keyboard is 42 _____ wide.

---

2. Use an inch ruler. What is the length of the string to the nearest inch?

Circle the number in the box to make the sentence true.

The string is
| 2 |
| 3 | inches long.
| 4 |

GO ON

3. Measure the crayon in centimeters and in inches.

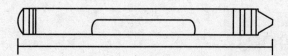

Circle the number or words in each box
that makes the sentence true.

The crayon is
| 6 |
| 7 |
| 8 |
centimeters long.

The crayon is about
| 1 |
| 2 |
| 3 |
inches long.

There are more
| inches than centimeters |
| centimeters than inches |
because
| inches are smaller than centimeters. |
| centimeters are smaller than inches. |

1. The paintbrush is about 7 centimeters long. Gavin says the feather is about 8 centimeters long. Ray says the feather is about 5 centimeters long.

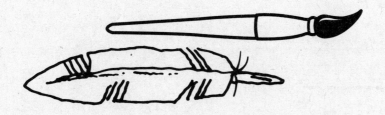

Which boy has the better estimate? Explain.

_____

_____

_____

2. Ella lays 1-inch tiles in a line until the line has the same length as a football.

Which sentence best estimates the length of a football?

○ A football is less than 6 tiles long.

○ A football is about 12 tiles long.

○ A football is more than 6 feet long.

○ A football is more than 50 tiles long.

GO ON ➡

3. The paper clip is about 4 centimeters long.
Robin says the string is about 7 centimeters
long. Gale says the string is about
20 centimeters long.

_____

Which girl has the better estimate? Explain.

_____

_____

4. Estimate the length of a real horse. Fill in the
bubble next to all the sentences that are true.

○ The horse is less than 1 meter long.

○ The horse is less than 6 meters long.

○ The horse is more than 3 centimeters long.

○ The horse is about 6 centimeters long.

○ The horse is about 3 meters long.

1. Alberto uses 8 centimeters of wire for a science project. He uses another 15 centimeters of wire for another project. How many centimeters of wire does he use?

   Draw a diagram. Write a number sentence using a ■ for the missing number. Then solve.

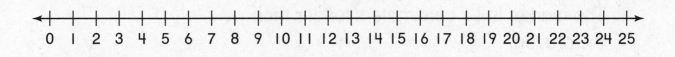

   _____

   Alberto uses _____ centimeters of wire.

2. Measure the length of each object. Choose all the sentences that describe the objects.

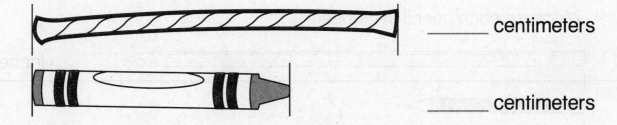

   _____ centimeters

   _____ centimeters

   ○ The yarn is 3 centimeters longer than the crayon.

   ○ The crayon is 7 centimeters shorter than the yarn.

   ○ The total length of the yarn and the crayon is 17 centimeters.

GO ON

3. Measure the length of each object. Choose all the sentences that describe the objects.

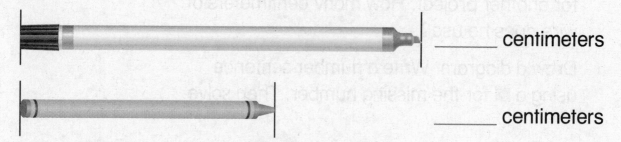

_____ centimeters

_____ centimeters

○ The marker is 11 centimeters longer than the crayon.

○ The crayon is 4 centimeters shorter than the marker.

○ The total length of the marker and the crayon is 18 centimeters.

4. Measure each pencil in inches.

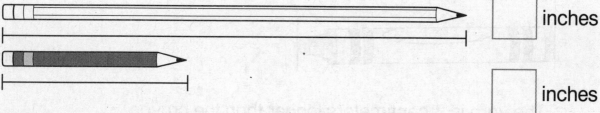

☐ inches

☐ inches

Which sentence is true?

○ The gray pencil is 4 inches longer than the white pencil.

○ The white pencil is 6 inches longer than the gray pencil.

○ The white pencil is 4 inches longer than the gray pencil.

1. Meg has a ribbon that is 9 inches long. She has another ribbon that is 11 inches long. How many inches of ribbon does Meg have?

Draw a diagram. Write a number sentence using a ■ for the missing number. Solve.

Meg has _____ inches of ribbon.

2. Michael has 54 centimeters of brown ribbon. He has 16 more centimeters of yellow ribbon than brown ribbon. How many centimeters of yellow ribbon does Michael have?

Michael has | 38
             | 60
             | 70 | centimeters of yellow ribbon.

3. Margaret wants to build a block tower 80 inches tall. So far her tower is 22 inches tall. How much taller does Margaret need to build her tower?

Margaret needs to build her tower | 52
                                   | 58
                                   | 102 | inches taller.

GO ON

4. Luke has a string that is 6 inches long and a string that is 11 inches long. How many inches of string does Luke have?

Draw a diagram. Write a number sentence using a ■ for the missing number. Solve.

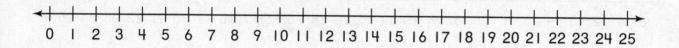

_____

Luke has _____ inches of string.

___

Write an equation and solve.

5. Mrs. Samson puts a fence around her patio. The patio has three sides. One side is 14 feet long. The second side is 17 feet long. The third side is 9 feet long. How many feet of fencing does Mrs. Samson use?

_____ = [   ] _____

unit

___

6. Olivia puts lace around the outside of a square pillow. Each side of the pillow is 18 centimeters long. How many centimeters of lace does Olivia need?

_____ = [   ] _____

unit

Practice Test

2.MD.B.6
Relate addition and subtraction to length.

1. Elizabeth has a piece of ribbon that is 25 centimeters long. She cuts off a piece of the ribbon to use to wrap a gift. Elizabeth's ribbon is now 7 centimeters long. How many centimeters of ribbon did Elizabeth use to wrap the gift?

Write a number sentence using a ■ for the missing number. Then solve.

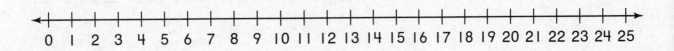

$$0 \quad 1 \quad 2 \quad 3 \quad 4 \quad 5 \quad 6 \quad 7 \quad 8 \quad 9 \quad 10 \quad 11 \quad 12 \quad 13 \quad 14 \quad 15 \quad 16 \quad 17 \quad 18 \quad 19 \quad 20 \quad 21 \quad 22 \quad 23 \quad 24 \quad 25$$

_____

Elizabeth used _____ centimeters of ribbon.

2. Complete the equation that the number line diagram represents.

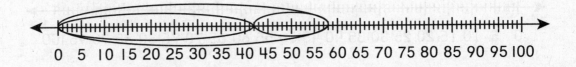

$$0 \quad 5 \quad 10 \quad 15 \quad 20 \quad 25 \quad 30 \quad 35 \quad 40 \quad 45 \quad 50 \quad 55 \quad 60 \quad 65 \quad 70 \quad 75 \quad 80 \quad 85 \quad 90 \quad 95 \quad 100$$

$$\boxed{} + 16 = \boxed{}$$

GO ON

3. Ethan's rope is 25 centimeters long. Ethan cuts the rope and gives a piece to Hank. Ethan's rope is now 16 centimeters long. How many centimeters of rope does Hank have?

Draw a diagram. Write a number sentence using a for the missing number. Then solve.

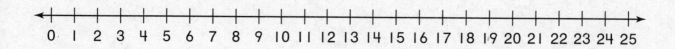

0 1 2 3 4 5 6 7 8 9 10 11 12 13 14 15 16 17 18 19 20 21 22 23 24 25

Hank has _____ centimeters of rope.

4. Represent the equation on the number line diagram.

$$47 + \blacksquare = 68$$

0   5   10 15 20 25 30 35 40 45 50 55 60 65 70 75 80 85 90 95 100

Find the difference.

$$47 + \boxed{} = 68$$

I. The clock shows the time Sarah starts
getting ready for bed. Write the time.
Circle A.M. or P.M.

_____

A.M.

P.M.

Tell how you knew whether to select A.M. or P.M.

_____

_____

2. Write the times the clocks show.

_____   _____   _____

GO ON ➡

Name _____

3. Write the times the clocks show.

_____ _____ _____

4. Write the times the clocks show.

_____ _____ _____

5. What time is shown on the clock? Fill in the bubbles next to all the ways to write or say the time.

○ 3:45

○ 9:15

○ 15 minutes past 9

○ half past 9

Practice Test
2.MD.C.8
Work with time and money.

1. Ben has 30¢. Circle coins to show this amount.

2. Marta pays $1.80 for a snack.

Fill in the bubble next to all the ways that
show $1.80.

○ one $1 bill, 2 quarters, and 3 dimes

○ one $1 bill, 3 quarters, and 1 nickel

○ 5 quarters and 3 dimes

○ 7 quarters and 1 nickel

3. Antoine gave Fran these coins. Antoine says he
gave Fran $1.00. Is Antoine correct? Explain.

_____

GO ON ➡

Name _____

4. James paid for a drink with this money.

Circle the amount to complete the sentence.

James paid a total of $\boxed{\begin{array}{c}\$1.41\\[2pt]\$1.46\\[2pt]\$1.61\end{array}}$ for the drink.

5. Hannah gave her sister these coins. Write the value of the coins. Explain how you found the total value.

_____

_____

6. Kim has 60¢. Circle coins to show this amount.

1. Drew made a line plot to show the lengths of his toy boats. How many boats are shown in the line plot?

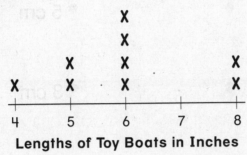

Lengths of Toy Boats in Inches

The line plot shows _____ toy boats.

Suppose one of the toy boats that is 5 inches long breaks and is thrown away. Explain how Drew should change the line plot.

_____

2. Dalia made a line plot to show the lengths of her ribbons. How many ribbons are shown in the line plot?

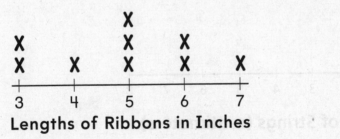

Lengths of Ribbons in Inches

The line plot shows _____ ribbons.

Suppose Dalia cut one of the ribbons that is 6 inches long into two pieces that are each 3 inches long. Explain how she should change the line plot.

_____

_____

GO ON

Name _____

3. Measure each string to the nearest centimeter. Then match the string to its length.

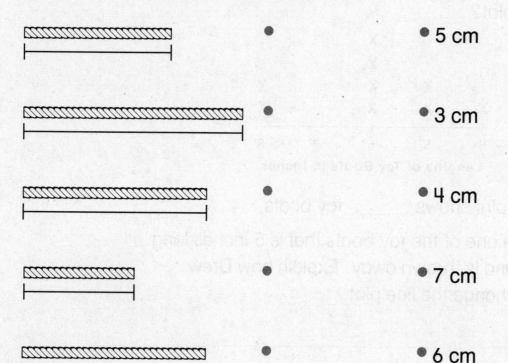

● 5 cm

● 3 cm

● 4 cm

● 7 cm

● 6 cm

4. Show the lengths of the five strings in question 3 on this line plot.

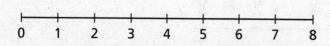

0 1 2 3 4 5 6 7 8

Lengths of Strings (centimeters)

How many strings are longer than 4 cm?

[ ] strings

1. Use the tally chart to complete the picture graph.
   Draw a ☺ for each child.

Favorite School Subject	
math	||||
reading	|||
science	||
art	|||

Favorite School Subject					
math					
reading					
science					
art					

Key: Each ☺ stands for 1 child.

2. How many children chose art?

   _____ children

3. How many more children chose math than
   science?

   _____ more children

4. Name two subjects that were chosen by a total of
   6 children.

   _____

GO ON ➤

Name _____

5. Eric asked some friends to name their favorite sandwiches. Use the data to complete the bar graph.

4 children like ham.

6 children like turkey.

1 child likes cheese.

3 children like jelly.

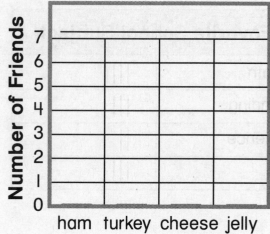

_____

6. Fill in the bubble next to all the sentences that describe the data in the bar graph above.

○ 6 children chose ham.

○ Turkey was the most popular.

○ 7 children chose cheese or jelly.

○ 5 more children chose turkey than cheese.

7. Did more children choose cheese or jelly than ham? Explain.

_____

_____

8. How many children chose a sandwich other than turkey?

_____ children

Name _____

1. Match the shapes.

  •   •

  •   •

  •   •

  •   •

2. Write the numbers that make each sentence true.

A rectangular prism has _____ faces.

Each face of a rectangular prism has _____ edges.

A rectangular prism has _____ vertices.

A rectangular prism has _____ edges.

GO ON

Name _____

3. Alex built this rectangular prism. Circle the number of unit cubes Alex used.

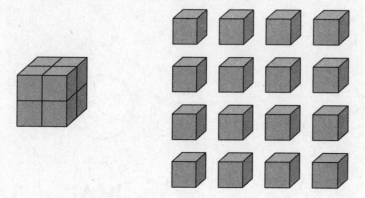

4. Paul makes a hexagon and a triangle with straws. He uses one straw for each side of a shape. How many straws does Paul need?

_____ straws

5. Draw each shape where it belongs in the chart.

Shapes with 4 or Fewer Angles	Shapes with More than 4 Angles

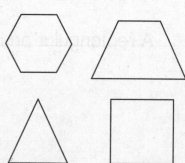

1. Max wants to cover the rectangle with gray tiles. Explain how you would estimate the number of gray tiles he would need to cover the rectangle.

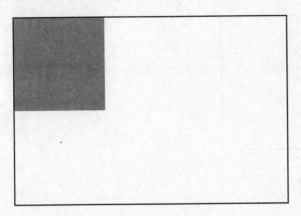

_____

_____

2. Measure in centimeters. Draw rows and columns. Write the number of small squares.

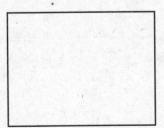

_____ squares

3. Measure in inches. Draw rows and columns. Write the number of small squares.

_____ squares

GO ON

Name _____

4. Grace wants to cover the rectangle with gray tiles. Explain how you would estimate the number of gray tiles she would need to cover the rectangle.

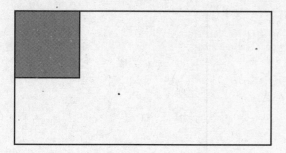

_____

_____

---

5. Mary started to cover this rectangle with ones blocks. Explain how you would estimate the number of ones blocks that would cover the whole rectangle.

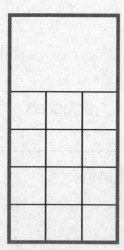

_____

_____

_____

STOP

Practice Test

2.G.A.3
Reason with shapes and their attributes.

1. Draw lines to show fourths.

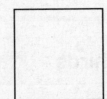

Explain how you know that the parts are fourths.

_____

_____

2. Hector makes two equal sized sandwiches. He cuts both sandwiches into thirds. What are two different ways he can cut the sandwiches? Draw to show your answer.

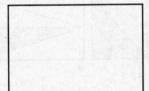
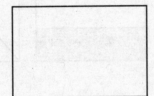

3. Fill in the bubble next to the shapes that show thirds.

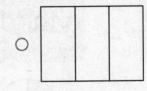

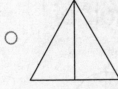
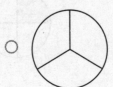

GO ON ▶

4. Draw lines in each shape to make equal shares.

Two Halves	Three Thirds	Four Fourths
□	□	□
○	○	○

5. Choose the rectangles that show a fourth shaded.

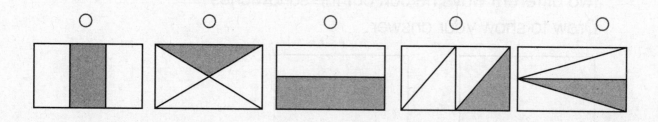

○   ○   ○   ○   ○

6. Write a term from a tile to tell how much is shaded.

a fourth of	a third of	a half of

STOP

1. Carlos uses $400 - 300$ to estimate how many more girls than boys are in a school.

   Fill in the bubble next to all the problems he may have been estimating for.

   ○ $\begin{array}{r} 395 \\ -218 \end{array}$ ○ $\begin{array}{r} 360 \\ -324 \end{array}$ ○ $\begin{array}{r} 480 \\ -341 \end{array}$ ○ $\begin{array}{r} 417 \\ -288 \end{array}$

2. Complete the equation that the diagram represents.

   0 5 10 15 20 25 30 35 40 45 50 55 60 65 70 75 80 85 90 95 100

   $\boxed{} + 16 = \boxed{}$

3. Use the data in the list to complete the line plot.

   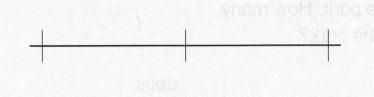

Lengths of Ribbons
6 inches
5 inches
7 inches
6 inches

4. Draw to show halves, thirds, and fourths.

   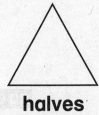   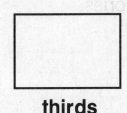   

   halves   **thirds**   **fourths**

   GO ON ➡

5. Draw and label coins to show a total value of 65¢.

---

6. Tricia has 12 pencils to share equally with her classmate. Draw a picture to show how Tricia can share her pencils.

How many pencils will Tricia keep?

_____ pencils

---

7. Kate sees 4 white dogs, 9 brown dogs, and 3 black dogs at the park. How many dogs does she see at the park?

_____ dogs

---

8. Aya has 167 baseball cards. She gives away 35 of her cards to a friend. Choose which step will help Aya to figure out how many baseball cards she has left.

○ Ungroup 7 tens as 6 tens 10 ones.

○ Subtract 5 ones from 6 tens.

○ Subtract 3 tens from 6 tens.

GO ON

9. Use an inch ruler. What is the length of the crayon to the nearest inch?

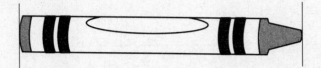

_____ inches

---

10. Use the 1-inch mark. Estimate the length of each piece of yarn.

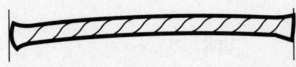

about _____ inches

about _____ inches

---

11. Measure in centimeters. Draw rows and columns. Write the number of small squares.

_____ squares

12. Frank counts by twos to 20. Elsie counts
by ones to 10. Who will say more numbers?
Explain.

_____

_____

---

13. Write an equation and solve.

Mr. Quito has a square pool. Each side of the pool is
16 feet long. How many feet is it around the whole pool?

_____ = [   ] _____
　　　　　　　　　　　　　unit

---

14. Use the words on the tiles to make the sentence true.

The boy is 40 _____ tall.

The car is 12 _____ long.

The driveway is 35 _____ long.

| inches | feet |

---

15. Shade in the ten frames to show the number.
Circle **even** or **odd**.

15

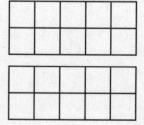

even　　　**odd**

16. Andy is thinking of a number that has a digit less than 5 in the tens place. It has a digit greater than 7 in the ones place. Fill in the bubble next to all the numbers that could be Andy's number.

○ 70 + 8

○ forty-nine

○ 2 tens 8 ones

Write another number Andy could be thinking of.

_____

17. Write a symbol from a tile to compare numbers.

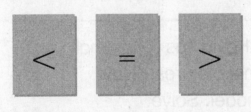

128 ◯ 137

18. Find the difference.

70
− 38

Tens	Ones
○ 2	○ 2
○ 3	○ 3
○ 4	○ 4

GO ON

19. Mason drew 2 two-dimensional shapes that had 8 angles in all. Draw the shapes Mason could have drawn.

20. A store sells 2 boxes of 100 pencils and some single pencils. Choose all the numbers that show how many pencils the store could sell.

   ○ 219   ○ 206

   ○ 120   ○ 182

21. Dan has 29 animal pictures. Kayla has 37 animal pictures. Who has more animal pictures? How many more? Label the bar model. Solve.

   Circle the word or number from each box to make the sentence true.

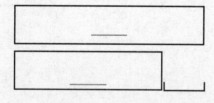

Dan	has	22	more animal pictures.
Kayla		12	
		8	

GO ON

22. Write the next number in each counting pattern.

338, 348, 358, 368, _____

472, 572, 672, 772, _____

---

23. Write the time that is shown on this clock.

_____ : _____

---

24. Describe how the number of sit-ups that Karina
did changed from Monday to Thursday. Make a
bar graph to solve the problem.

Monday—3 sit-ups   Wednesday—6 sit-ups

Tuesday—5 sit-ups   Thursday—8 sit-ups

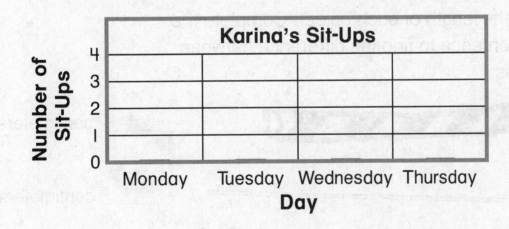

_____

_____

GO ON ➡

25. Write how many tens. Circle groups of 10 tens.
Write how many hundreds. Write the number.

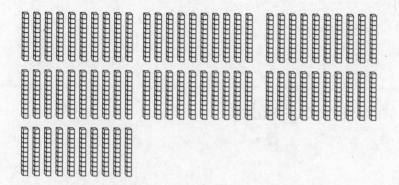

_____ tens

_____ hundreds

_____

26. Mandy has 26 baseball cards, 42 football cards,
and 38 basketball cards. How many cards does
she have?

| 68 |
| 80 |
| 106 |

Mandy has [80] cards.

27. Measure the length of each object. Complete the
number sentence to find the difference between
the lengths.

_____ centimeters

_____ centimeters

_____ − _____ = _____

The string is _____ centimeters longer than the straw.

GO ON ➡

Two Schools

**Jefferson School has students in 1st grade
to 5th grade.**

1. The number of children in 1st grade has 3 digits.
   The digits in the number are 2, 3, and 8.
   The digit 8 means 80 in this number.
   Write a number that could be the number of
   children in 1st grade.

2. Write a number that is 10 less than the number of
   children you chose for 1st grade.

3. Write a number sentence that uses $>$, $<$, or $=$
   to compare your answers from questions 1 and 2.

GO ON

4. Donell uses these blocks to show the number of students in 3rd grade.

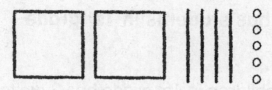

How many students are in 3rd grade?

_____ students

5. There are 100 more students in 4th grade than in 5th grade. Grade 5 has 176 students.
Draw a quick picture to show how many students are in 4th grade.

6. Write a number sentence that uses >, <, or = to compare the number of students in 4th grade with the number of students in 3rd grade.

_____ ◯ _____

**Yasmeen goes to Lincoln School. She counts the
number of 2nd grade students who go there.
The number in the circle is the total number of
2nd grade students at Lincoln School.**

7. Fill in the missing numbers. Count by tens.

220, 230, _____, _____, _____, _____, _____

8. Yasmeen uses tens blocks to show the number of
2nd grade students. How many tens blocks will she need?

She will need _____ blocks.

9. Suppose Yasmeen's school has 2 I 0 students in
3rd grade. How would you figure out a number
that is I 0 more than that? Write your answer.
Explain how you know.

_____

_____

_____

_____

GO ON ➡

**The number of students at Jefferson School
is even. The number has three digits.
The digit in the tens place is 4.**

10. Write three numbers that could be the number of
students at Jefferson School.

_____   _____   _____

Explain how you know your answers are correct.

_____

_____

_____

11. Choose one of the numbers that you just wrote.
Write it three different ways.

12. Write a 3-digit number that could NOT be the
number of students at Jefferson School.

There could NOT be _____ students.

1. Sally scores more points in a game than Ty.
Sally uses 900 − 500 to estimate how many more points.

Fill in the bubble next to all the problems she may
have been estimating for.

○ 892
  − 502

○ 794
  − 499

○ 922
  − 598

○ 905
  − 510

2. Complete the equation that the diagram represents.

0   5   10 15 20 25 30 35 40 45 50 55 60 65 70 75 80 85 90 95 100

☐ + 17 = ☐

3. Use the data in the list to complete the line plot.

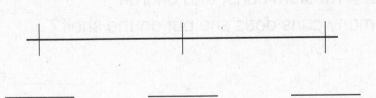

| **Lengths of Ribbons** |
| 7 inches |
| 5 inches |
| 7 inches |
| 6 inches |

4. Draw to show halves, thirds, and fourths.

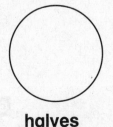

halves

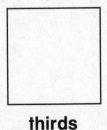

thirds

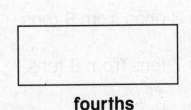

fourths

GO ON

5. Draw and label coins to show a total value of 76¢.

___

6. Draw a picture to show how Jose can share
18 strawberries with his brother equally.

How many strawberries will Jose receive?

_____ strawberries

___

7. Erin puts 3 small cans, 4 medium cans, and 5 large
cans on a shelf. How many cans does she put on the shelf?

_____ cans

___

8. Lara subtracts 73 from 188. Which one of these
steps should she follow?

○ Ungroup 8 tens as 7 tens 10 ones.

○ Subtract 3 ones from 8 tens.

○ Subtract 7 tens from 8 tens.

GO ON ➤

9. Use an inch ruler. What is the length of the paper clip to the nearest inch?

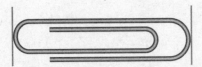

_____ inches

---

10. Use the 1-inch mark. Estimate the length of each object.

|← 1 inch →|

about _____ inches

about _____ inches

---

11. Measure in centimeters. Draw rows and columns. Write the number of small squares.

_____ squares

GO ON ➡

12. Megan counts by ones to 10. Lee counts by fives to 20. Who will say more numbers? Explain.

_____

_____

_____

13. Mr. Polley puts a fence around his yard. The yard has three sides. One side is 12 feet long. The second side is 7 feet long. The third side is 15 feet long. How many feet of fence does he need?

_____ = ☐ _____
                              unit

14. Use the words on the tiles to make the sentence true.

The table is 3 _____ long.

The belt is 30 _____ long.

The hallway is 15 _____ long.

| inches | feet |

15. Shade in the ten frames to show the number.
Circle **even** or **odd**.

18

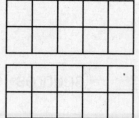

even   odd

16. Ann has a favorite number. It has a digit less than 4 in the tens place. It has a digit greater than 6 in the ones place. Fill in the bubble next to all the numbers that could be Ann's number.

○ 30 + 9

○ sixty-seven

○ 2 tens 8 ones

Write another number that could be Ann's favorite.

17. Write a symbol from a tile to compare numbers.

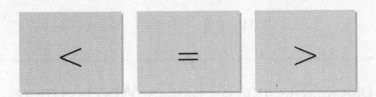

152 ◯ 142

18. Find the difference.

90
− 62

Tens	Ones
○ 2	○ 1
○ 3	○ 2
○ 5	○ 8

GO ON

19. Kevin drew 2 two-dimensional shapes that had
9 angles in all. Draw the shapes Kevin could
have drawn.

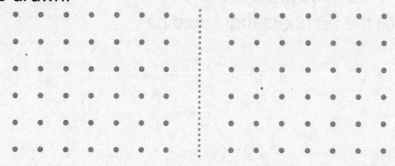

20. There are 4 boxes of 100 sheets of paper and some single
sheets of paper in the closet. Choose all the numbers
that show how many sheets of paper could be in the closet.

○ 348   ○ 324

○ 406   ○ 411

21. Ed has 28 blocks. Sue has 34 blocks. Who has more
blocks? How many more? Label the bar model. Solve.

Circle the word and number from each box to
make the sentence true.

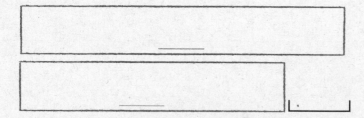

Ed		6	
	has	16	more blocks.
Sue		52	

GO ON

22. Write the next number in each counting pattern.

214, 314, 414, 514, _____

123, 133, 143, 153, _____

23. Write the time that is shown on this clock.

_____ : _____

24. Look at the bar graph. If Karina does 2 more sit-ups on Friday than she did on Thursday, how does the number of sit-ups change from Monday to Friday?

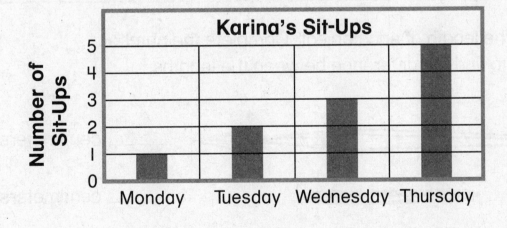

_____

_____

GO ON ➤

25. Write how many tens. Circle groups of 10 tens.
Write how many hundreds. Write the number.

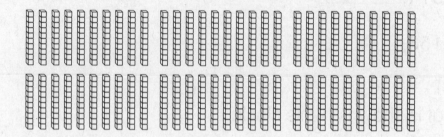

_____ tens

_____ hundreds

_____

26. Carlos has 23 red keys, 36 blue keys, and
44 green keys. How many keys does he have?

Carlos has
| 67 |
| 80 | keys.
| 103 |

27. Measure the length of each object. Complete the number
sentence to find the difference between the lengths.

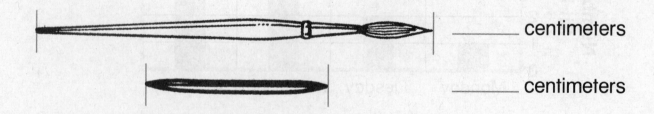

_____ centimeters

_____ centimeters

_____ − _____ = _____
centimeters centimeters centimeters

The paintbrush is _____ centimeters longer than
the toothpick.

The Reading Challenge

**Some children did a reading challenge.
They recorded how many books they read.**

1. Miguel read 7 books. Mia read the same number
   of books as Miguel. Write and solve the number
   sentence to show how many books Miguel and
   Mia read in all.

   _____ + _____ = _____

---

2. Abdul read 8 books. Jose read 11 books.
   How many more books did Jose read than Abdul?

   _____ books

---

3. Estela read 23 books last month. She put the
   books onto three shelves. Each shelf has a
   different number of books.

   Write a number sentence to show how many
   books might be on each shelf.

   _____ + _____ + _____ = _____

**Franco read 22 books. Tia read 15 books.
Ali read more books than Tia but fewer than
Franco.**

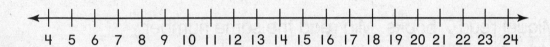

4. How many more books did Franco read than Tia?
   Use the number line to solve.

   _____ books

_____

5. How many books could the 3 children have read in all?
   Write a number sentence.

   _____ + _____ + _____ = _____

   Explain your answer.

   _____

   _____

   _____

GO ON

6. Erica read 31 books. Her friend Molly read
17 books. Draw a quick picture to show
how many books Erica and Molly read in all.

They read _____ books in all.

7. Serena read a book that is 38 pages long. Ming read
a book that is 26 pages long. Rohan read a book
that is 31 pages long.

Ming finds how many pages he and Rohan read in all.
How many more pages did they read than Serena?

They read _____ more pages than Serena.

Explain how you know your answer is correct.

_____

_____

_____

_____

GO ON ➡

**This tells how many books were read by
all the children in 3 different classes.**

Mr. Dorn's class	Ms. Lopez's class	Ms. Chen's class
319 books	185 books	241 books

8. How many books did the 3 classes read in all?

   _____ books

9. How many more books did Mr. Dorn's class read than
   Ms. Lopez's class?

   _____ books

10. The library has a shelf of storybooks. There are
    473 new books on the shelf. The children at the
    school have already read 205 of these books.
    How many of the books have they NOT read yet?

    _____ books

1. There are more fish than frogs in a pond. Zia uses 800 − 200 to estimate how many more fish.

   Fill in the bubble next to all the problems he may have been estimating for.

   ○ 775   ○ 814   ○ 891   ○ 811
      − 284   − 231   − 205   − 178

2. Complete the equation that the diagram represents.

   0 5 10 15 20 25 30 35 40 45 50 55 60 65 70 75 80 85 90 95 100

   ☐ + 24 = ☐

3. Use the data in the list to complete the line plot.

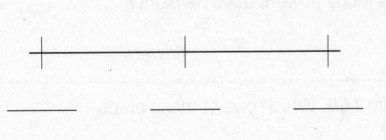

   _____ _____ _____

   _____

   Lengths of Ribbons

   6 inches
   5 inches
   7 inches
   5 inches

4. Draw to show halves, thirds, and fourths.

   halves   **thirds**   **fourths**

GO ON

5. Draw and label coins to show a total value of 87¢.

[ ]

---

6. Mark has 16 crackers to share equally with his mother. Draw a picture to show how Mark can share his crackers.

[ ]

How many crackers will Mark keep?

_____ crackers

---

7. Hank plants 4 white flowers, 7 pink flowers, and 3 yellow flowers. How many flowers does he plant?

_____ flowers

---

8. Fred subtracts 48 from 165. Which one of these steps should he follow first?

○ Ungroup 6 tens as 5 tens 10 ones.

○ Subtract 5 ones from 4 tens.

○ Subtract 5 ones from 8 ones.

GO ON ➤

9. Use an inch ruler. What is the length of the marker to the nearest inch?

_____ inches

10. Use the 1-inch mark. Estimate the length of each piece of yarn.

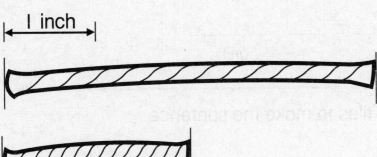

about _____ inches

about _____ inches

11. Measure in centimeters. Draw rows and columns. Write the number of small squares.

_____ squares

GO ON ➡

12. Arnod counts to 20 by fives. Bimi counts to 50 by tens. Who will say more numbers?

_____

_____

13. Sally puts a frame around a picture. The picture has four sides. Two sides are each 11 inches long. The other two sides are each 8 inches long. How many inches of frame does Sally need?

_____ = ☐ _____
unit

14. Use the words on the tiles to make the sentence true.

The tree is 18 _____ tall.

The desk is 30 _____ long.

The lamp is 3 _____ tall.

| inches | feet |

15. Shade in the ten frames to show the number. Circle **even** or **odd**.

20

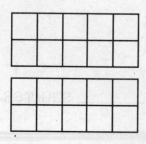

even   odd

GO ON ➡

16. Jack's age has a digit less than 4 in the tens place. It has a digit greater than 6 in the ones place. Which one of these numbers could be Jack's age?

○ 40 + 9

○ thirty-seven

○ 2 tens 5 ones

Write a number that could be Jack's age.

_____

17. Write a symbol from a tile to compare numbers.

136 ◯ 117

18. Find the difference.

   6 1
 − 3 7
 _____

Tens	Ones
○ 2	○ 2
○ 3	○ 3
○ 4	○ 4

19. Rana drew 2 two-dimensional shapes that had 10 angles in all. Draw the shapes Rana could have drawn.

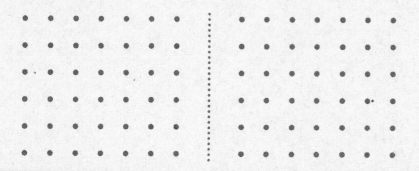

20. A store has 3 boxes of 100 juice bottles and some single bottles. Circle all the numbers that show how many bottles could be in the store.

○ 348   ○ 324

○ 406   ○ 411

21. Ed has 44 stickers. Marcy has 53 stickers. Who has more stickers?

Circle the word or number from each box to make the sentence true.

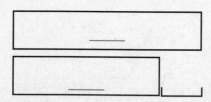

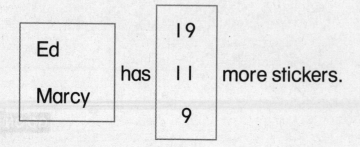

has   more stickers.

Ed	19
Marcy	11
	9

GO ON ➡

22. Write the next number in each counting pattern.

885, 875, 865, 855, _____

641, 541, 441, 341, _____

---

23. Write the time that is shown on this clock.

_____ : _____

---

24. Use the tally chart to complete the picture graph.
Draw a ☺ for each child.

Favorite School Subject	
math	||||
reading	|||
science	||
art	|||

Favorite School Subject				
math				
reading				
science				
art				

Key: Each ☺ stands for 1 child.

GO ON ➤

25. Write how many tens. Circle groups of 10 tens.
Write how many hundreds. Write the number.

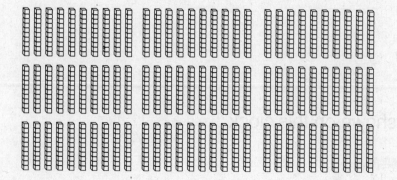

_____ tens

_____ hundreds

_____

26. Hana picks 35 red apples, 16 green apples, and
27 yellow apples. How many apples does Hana
pick?

Hana picks
| 65 |
| 78 | apples.
| 84 |

27. Measure the length of each object. Complete the
number sentence to find the difference between
the lengths.

_____ centimeters

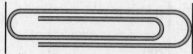

_____ centimeters

_____ − _____ = _____

The string is _____ centimeters longer than the paper clip.

The Museum Store

1. The museum store opens at 10:00 A.M.
   Eva gets to the store before 11:30 A.M.
   Draw hands on the clock to show when
   Eva might get to the store. Then write the time.

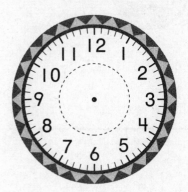

2. The clock shows when the store closes.
   Write the time.

Is this an A.M. time or a P.M. time? Explain how you know.

_____

_____

_____

3. Jin buys a bookmark.
   She uses these coins to pay for the bookmark.
   How much does she pay?

_____ ¢

---

4. Jin also buys a pencil.
   The pencil costs more than the bookmark.
   The pencil costs less than one dollar.

   How much could the pencil cost?

   The pencil could cost _____ ¢.
   Draw and label coins to show one way to make
   this amount.

---

5. Jin compares the lengths of her bookmark and her
   pencil. The pencil is about 12 cm long. What is a
   good estimate for the length of the bookmark?

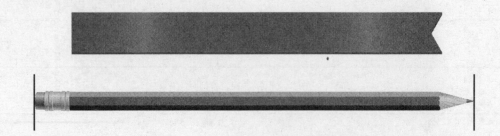

   The bookmark is about _____ cm long.

   GO ON ➡

The picture graph shows how many toy trucks of each color are in the store.

Colors of Trucks								
red	☺	☺	☺					
blue	☺	☺	☺	☺	☺			
black	☺	☺	☺	☺				
white	☺	☺	☺					

Key: Each ☺ stands for 1 truck.

---

6. How many black trucks are there? _____

---

7. Write color names to complete the sentence.

   The number of _____ trucks is equal

   to the number of _____ trucks.

---

8. The clerk sold 20 stuffed animals on Wednesday.
   He sold bears, lions, foxes, and turtles. Three of the
   animals were bears. Make a bar graph to show the
   stuffed animals.

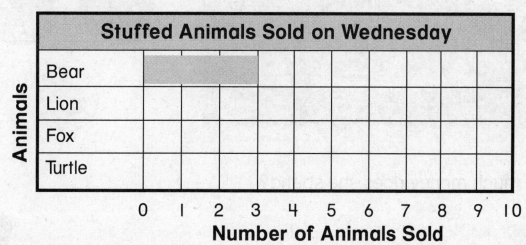

GO ON

9. Grace buys a black pen at the store.

Use an inch ruler. Measure the length of the pen
to the nearest inch.

The pen is about _____ inches long.

---

10. Grace also buys two banners for her room.
The first is 33 inches long. The second banner is 19 inches long.
How much longer is the first banner than the second banner?

_____ − _____ = _____

_____ inches longer

---

11. Grace uses this money to pay for her items.

How much money does she spend? _____

The Apartment Building

There is a big apartment building near the park.
Each apartment has a 3-digit number.
Jose's apartment number has the digit 9 in the
ones place and the digit 1 in the hundreds place.

1. Write a number that could be Jose's
   apartment number.

   _____

2. Erik lives in another apartment in the same building.
   The number of his apartment is 100 more than
   the number of Jose's apartment. What could
   Erik's apartment number be?

   _____

3. Marta lives in apartment 450. Write a number sentence
   that uses the symbols >, <, or = to compare
   Marta's apartment number and Erik's apartment number.

GO ON

Mila made these models to show how many apartments are in each tower of the building.

Tower D	Tower E

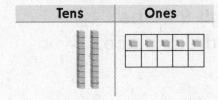 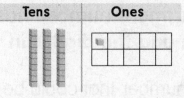

4. How many apartments are in Tower D?

_____ apartments

---

5. What is the total number of apartments in Towers D and E?

_____ apartments

---

6. Tower F will have an odd number of apartments less than 20. Pick the number. Then draw and explain to show that the number is odd.

Write the number.

_____ apartments

Raj moves into the building.

7. Raj unpacks some red plates and some
white plates. There are 30 plates in all.
Write how many red and white plates there could be.

_____ red plates   _____ white plates

---

8. Raj unpacks his books onto three shelves. He puts
between 100 and 150 books on each shelf.
He puts 10 more books on the bottom shelf
than on the middle shelf. He puts 10 fewer books
on the top shelf than on the middle shelf. Write the
number of books that Raj could put on each shelf.

_____

– – – – – – – – – – –   top shelf

_____

– – – – – – – – – – –   middle shelf

_____

– – – – – – – – – – –   bottom shelf

_____

How many books are there in all?

_____ + _____ + _____ = _____

There are _____ books.

GO ON ➡

9. Suppose the apartment building has 100 apartments
on each floor. Write the number of apartments
in the building for each number of floors.

5 floors | _____ apartments

6 floors | _____ apartments

9 floors | _____ apartments

2 floors | _____ apartments

8 floors | _____ apartments

10. Circle one of the problems in question 9. Draw and
write to show how you solved it.